Grid Paper

100 pages for all your graphing, designing, and drawing needs

Grid Paper: 100 pages for all your graphing, designing, and drawing needs

ISBN: 9798472118606

Cover design created by Jessica Lindie